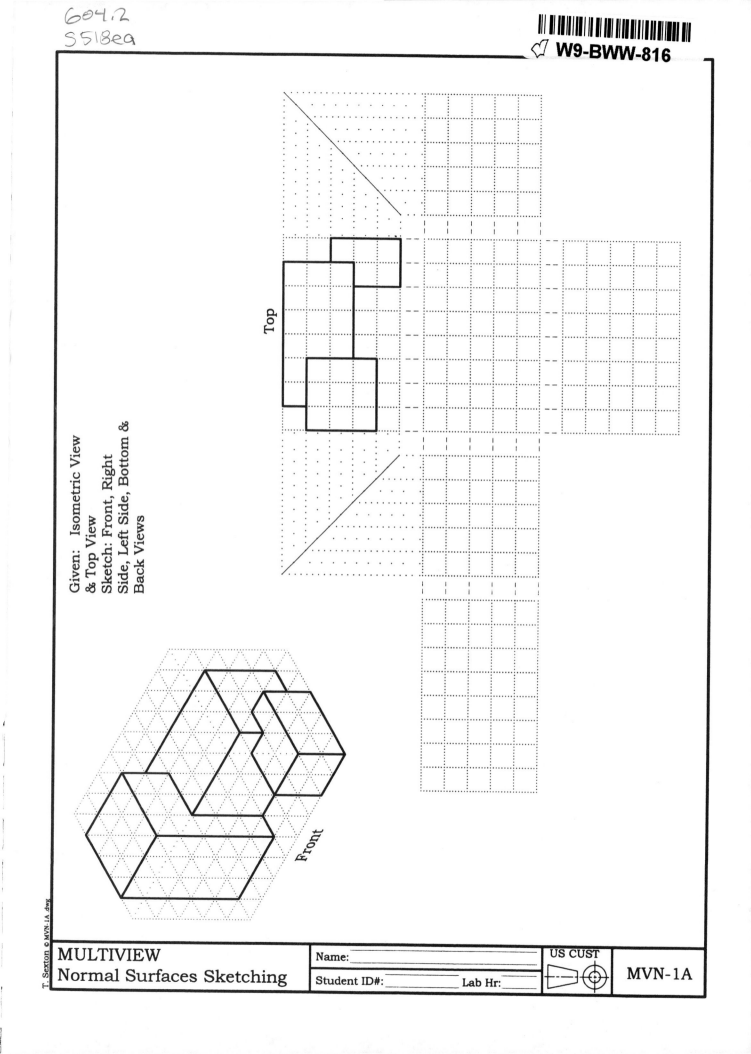

Given: Isometric View
& Top View
Sketch: Front, Right
Side, Left Side, Bottom &
Back Views

Top

Front

T. Sexton © MVN-1A.dwg

| MULTIVIEW | Name: | | US CUST | |
| Normal Surfaces Sketching | Student ID#: | Lab Hr: | | MVN-1A |

Given: Isometric View
Sketch: Front, Right
Side, Left Side, Bottom &
Back Views

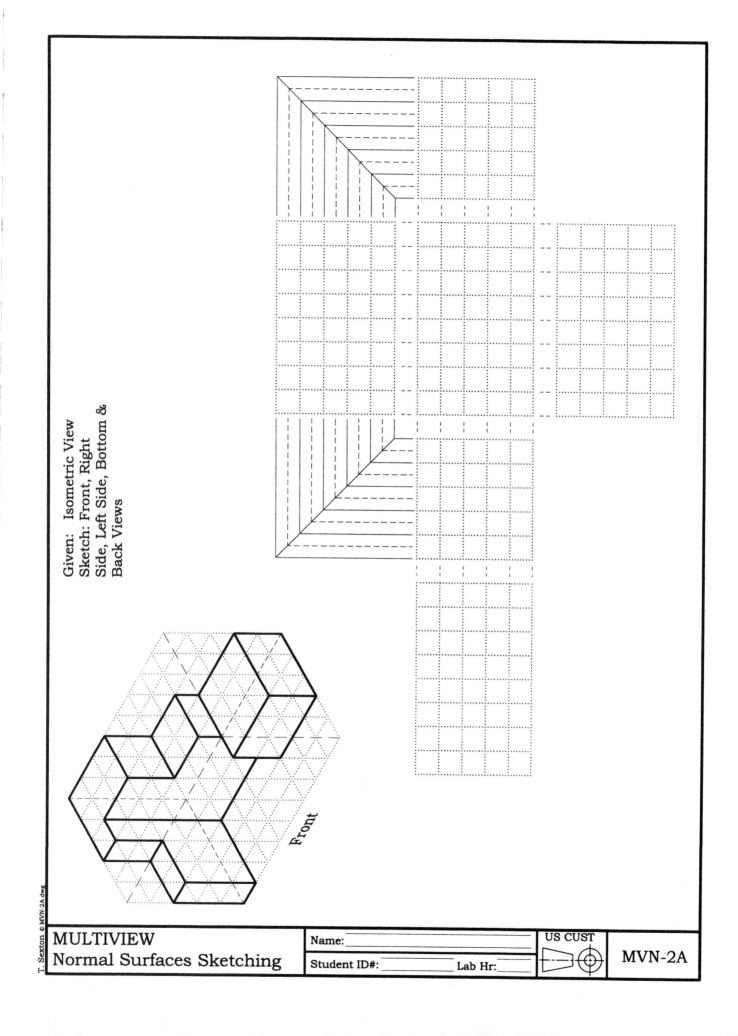

Front

| MULTIVIEW | Name: | US CUST | |
| Normal Surfaces Sketching | Student ID#: _____ Lab Hr: _____ | | MVN-2A |

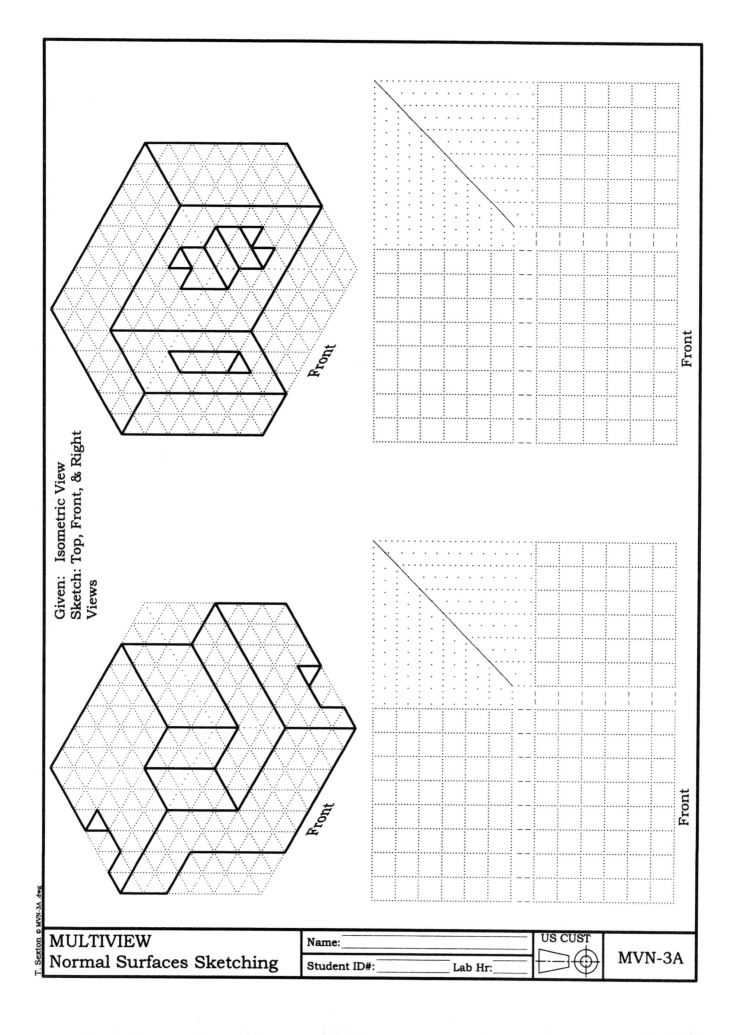

Given: Isometric View
Sketch: Top, Front, & Right
Views

Front

Front

Front

Front

MULTIVIEW
Normal Surfaces Sketching

Name:_____

Student ID#:_____ Lab Hr:_____

US CUST

MVN-3A

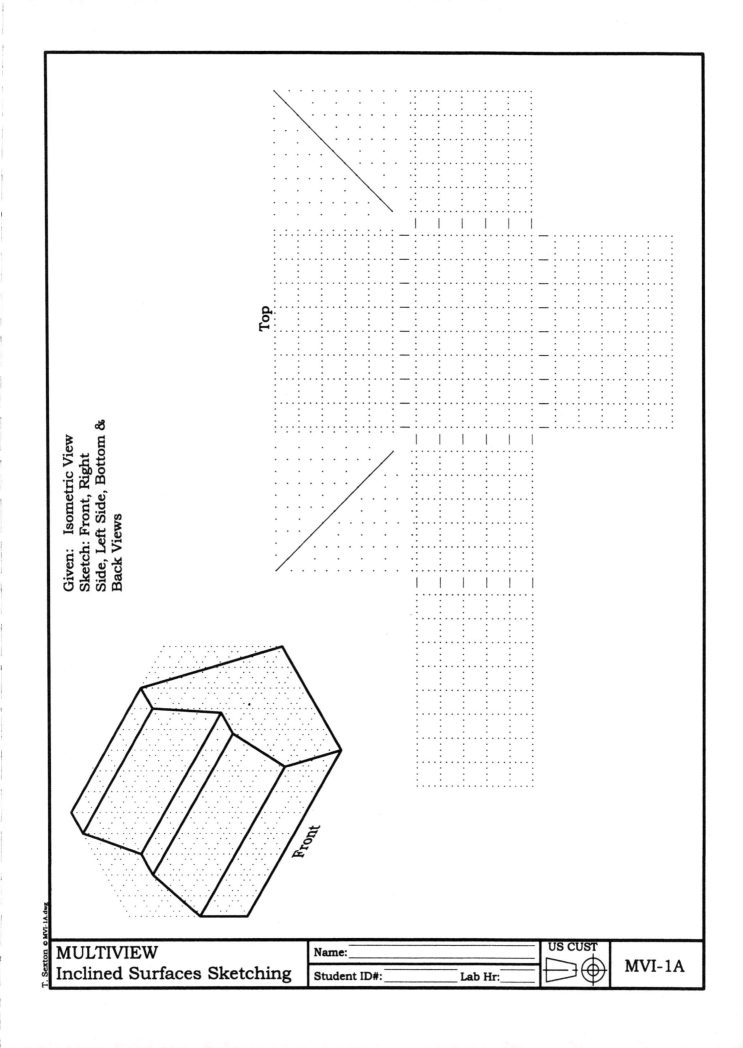

Given: Isometric View
Sketch: Front, Right
Side, Left Side, Bottom &
Back Views

Top

Front

| MULTIVIEW | Name: | | US CUST | |
| Inclined Surfaces Sketching | Student ID#: | Lab Hr: | | MVI-1A |

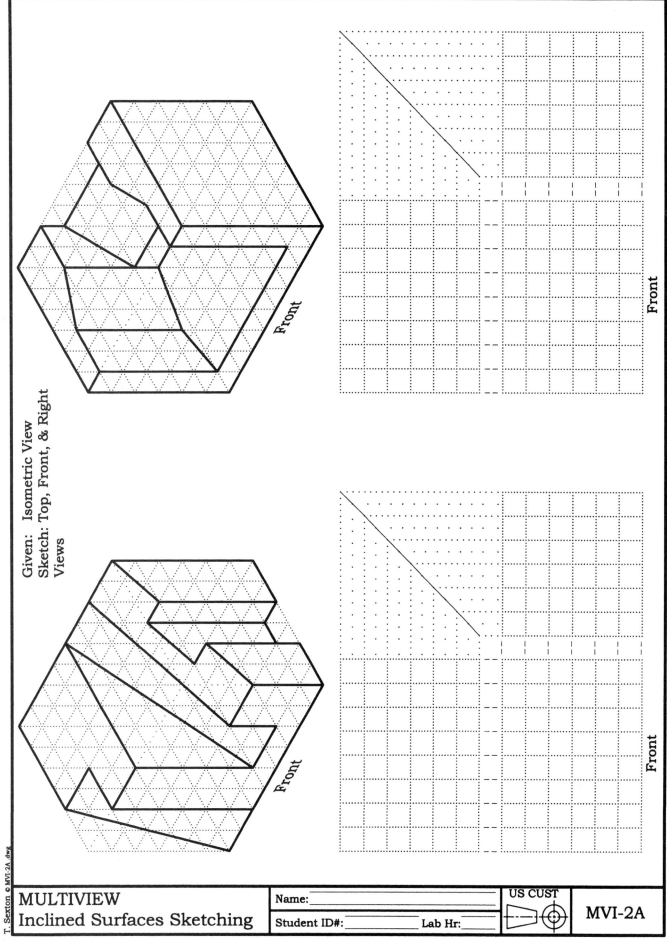

Given: Isometric View
Sketch: Top, Front, & Right
Views

Front

Front

Front

Front

MULTIVIEW
Inclined Surfaces Sketching

Name:_____

Student ID#:_____ Lab Hr:_____

US CUST

MVI-2A

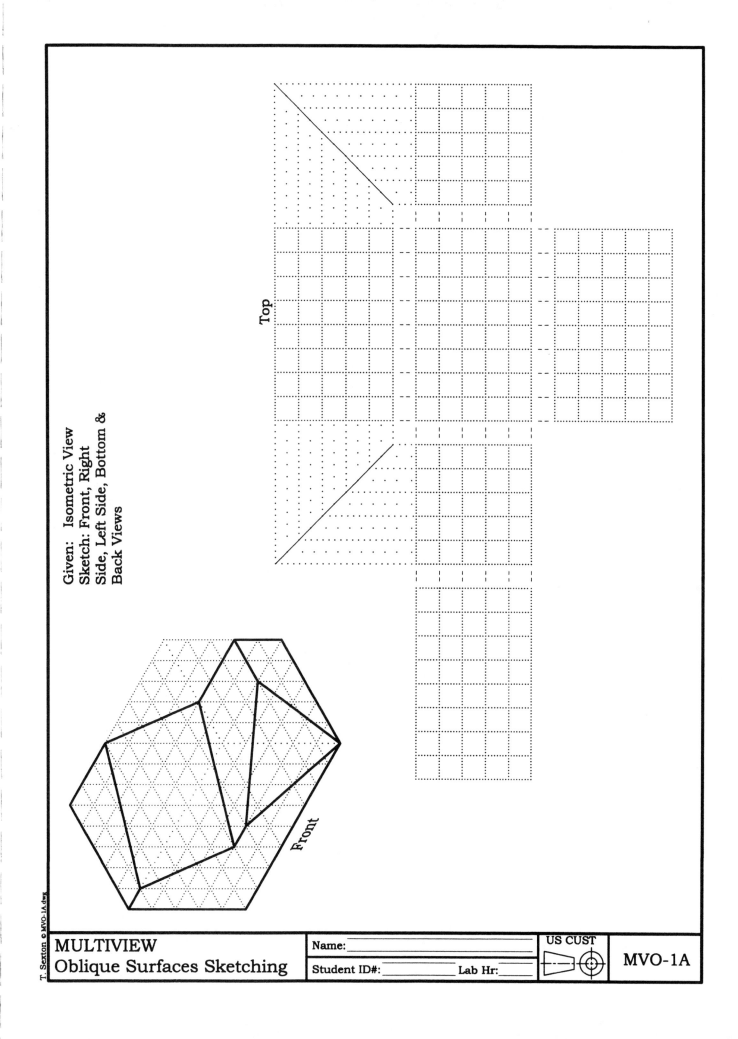

Given: Isometric View
Sketch: Front, Right
Side, Left Side, Bottom &
Back Views

Top

Front

MULTIVIEW
Oblique Surfaces Sketching

Name:_____

Student ID#:_____ Lab Hr:_____

US CUST

MVO-1A

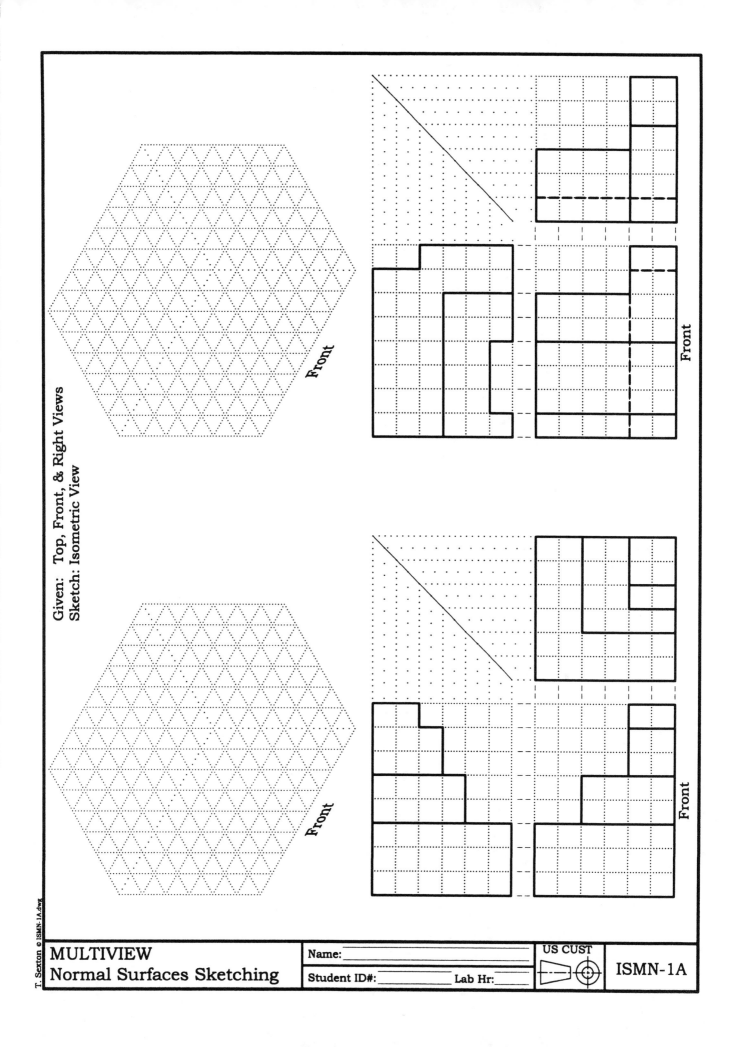

Given: Top, Front, & Right Views
Sketch: Isometric View

Front

Front

Front

Front

MULTIVIEW
Normal Surfaces Sketching

Name:

Student ID#:_____ Lab Hr:_____

US CUST

ISMN-1A

T. Sexton © ISMN-1A.dwg

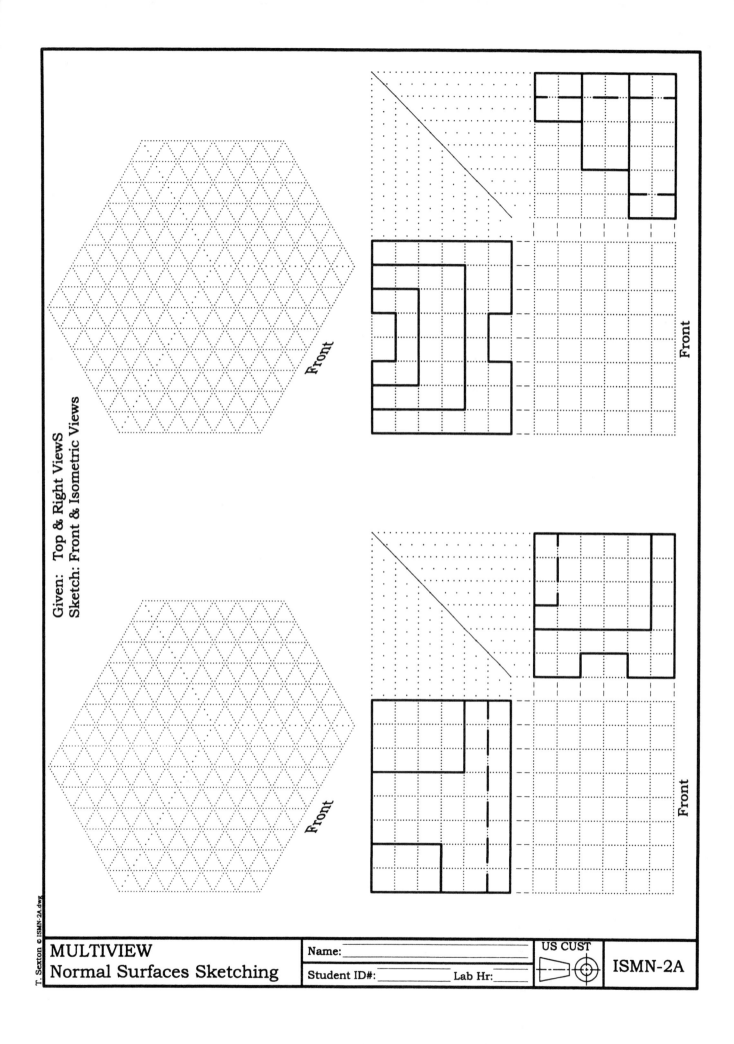

Given: Top & Right ViewS
Sketch: Front & Isometric Views

Front

Front

Front

Front

MULTIVIEW
Normal Surfaces Sketching

Name:_____

Student ID#:_____ Lab Hr:_____

US CUST

ISMN-2A

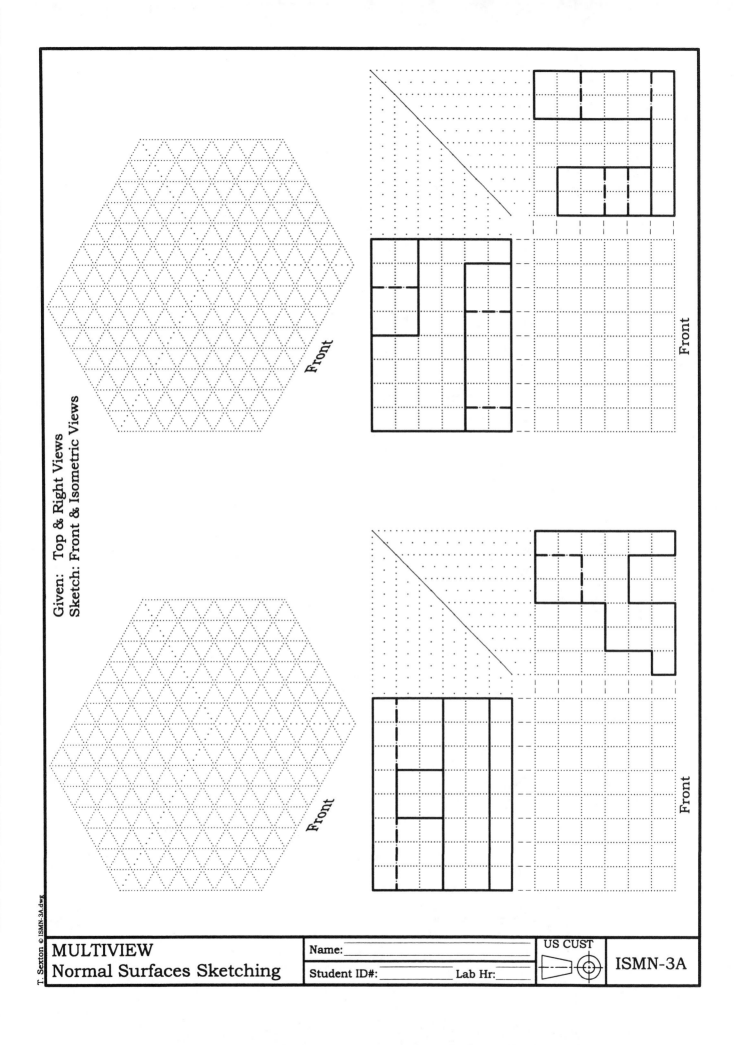

Given: Top & Right Views
Sketch: Front & Isometric Views

Front

Front

Front

Front

MULTIVIEW
Normal Surfaces Sketching

Name:

Student ID#: _____ Lab Hr: _____

US CUST

ISMN-3A

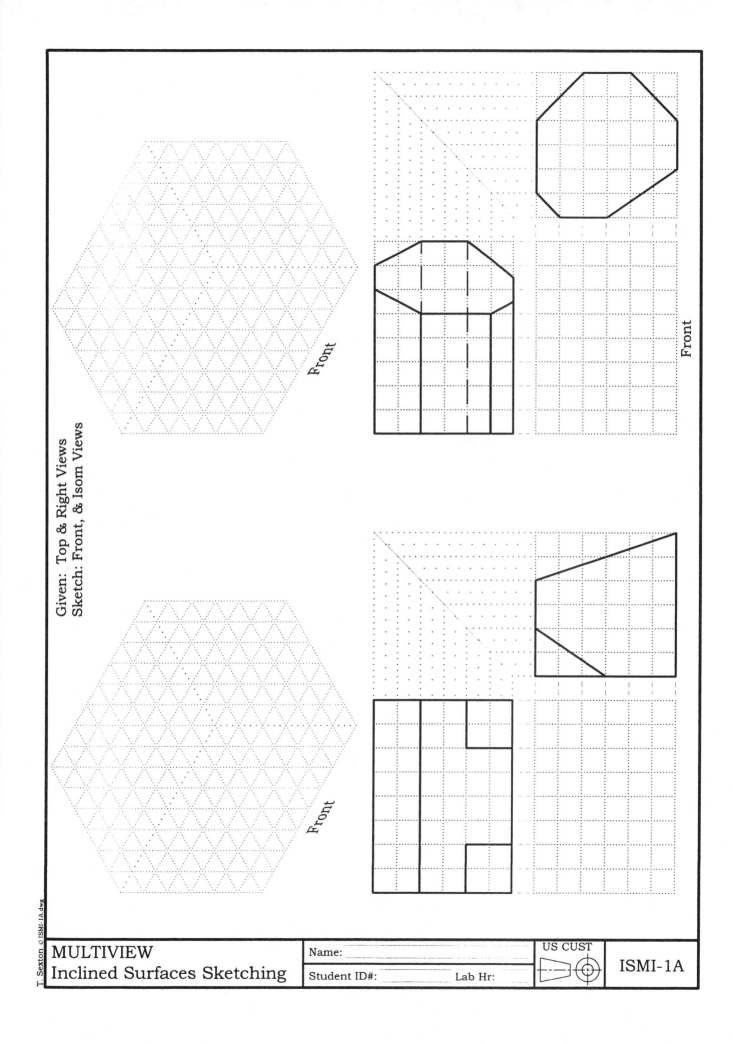

Given: Top & Right Views
Sketch: Front, & Isom Views

Front

Front

Front

Front

MULTIVIEW
Inclined Surfaces Sketching

Name:

Student ID#:   Lab Hr:

US CUST

ISMI-1A

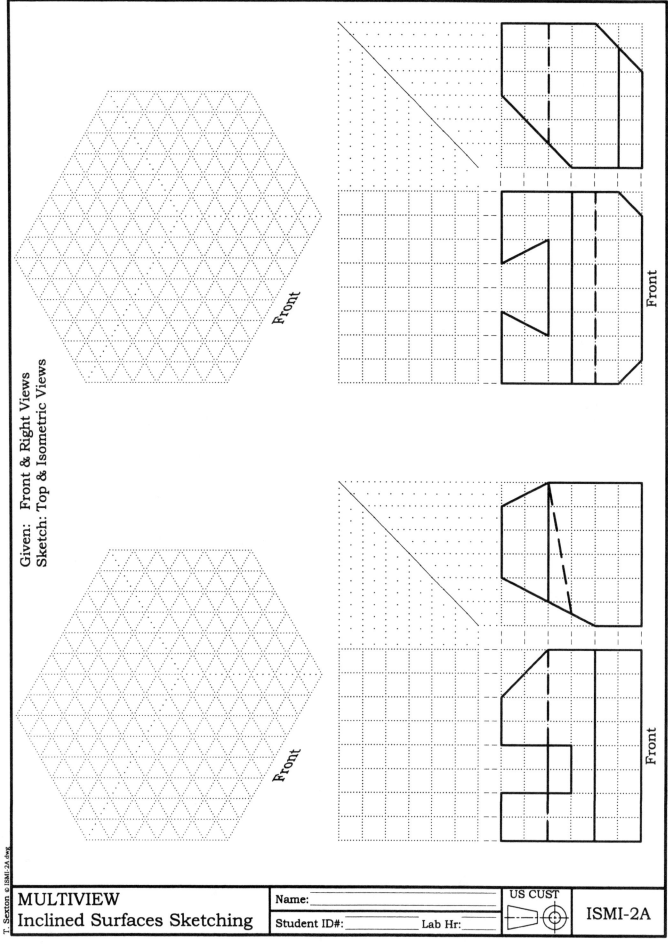

Given: Front & Right Views
Sketch: Top & Isometric Views

Front

Front

Front

Front

MULTIVIEW
Inclined Surfaces Sketching

Name:_____

Student ID#:_____ Lab Hr:_____

US CUST

ISMI-2A

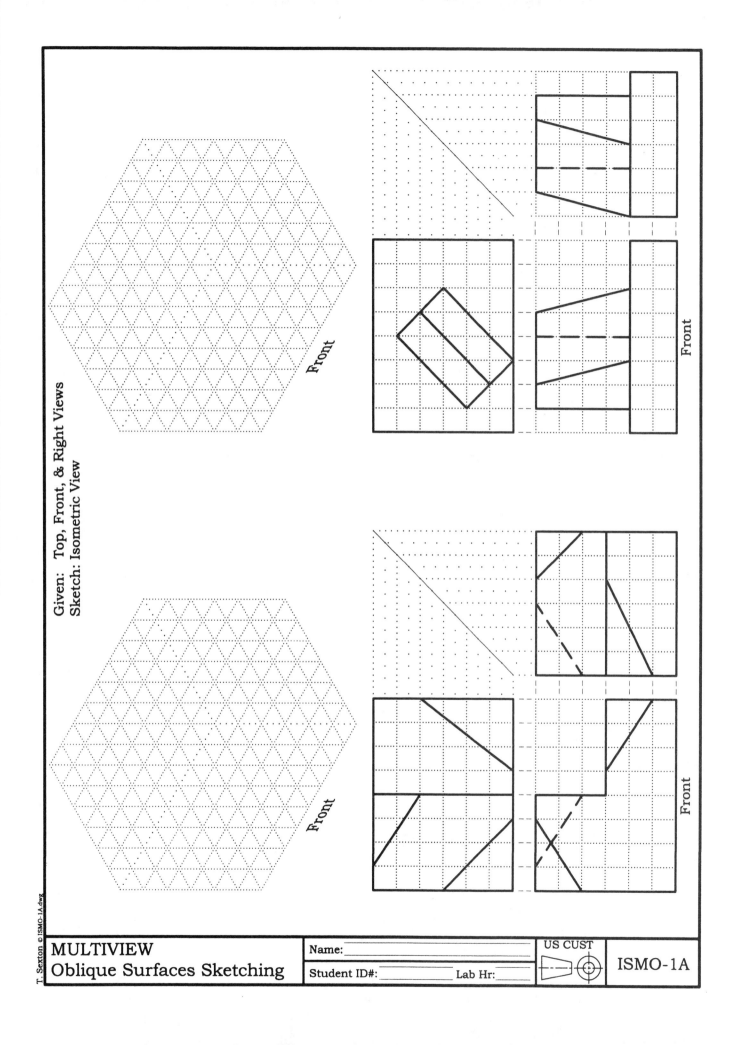

Given: Top, Front, & Right Views
Sketch: Isometric View

Front

Front

Front

Front

MULTIVIEW
Oblique Surfaces Sketching

Name:_____

Student ID#:_____ Lab Hr:_____

US CUST

ISMO-1A

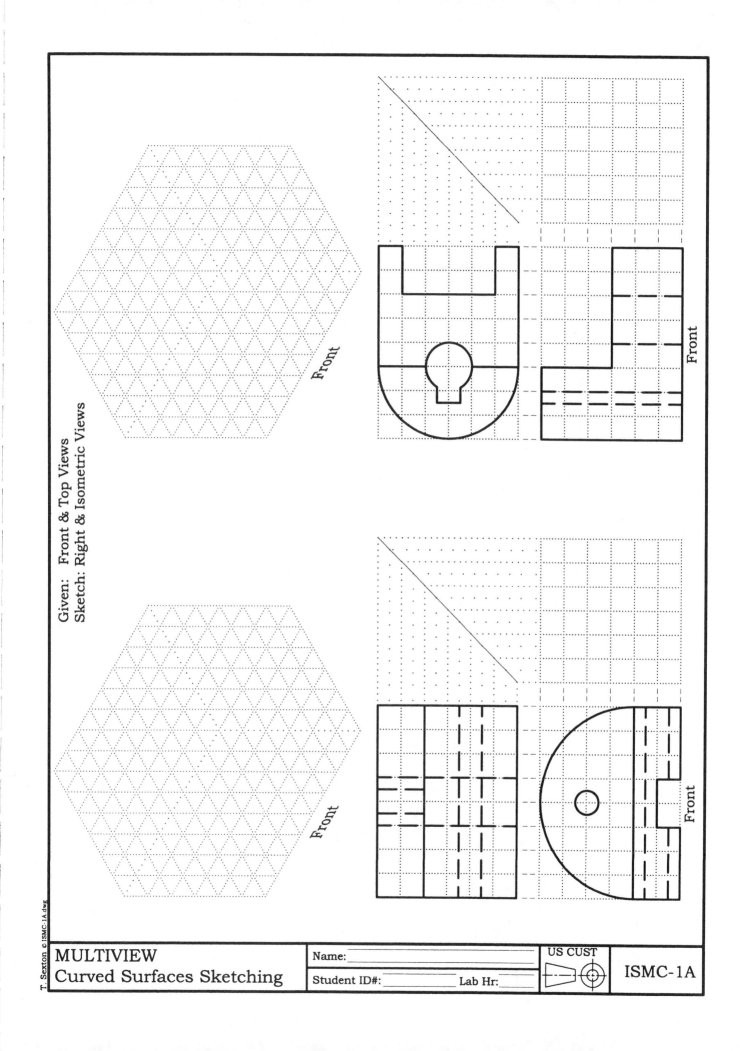

Given: Front & Top Views
Sketch: Right & Isometric Views

Front

Front

Front

Front

T. Sexton © ISMC-1A.dwg

MULTIVIEW
Curved Surfaces Sketching

Name:

Student ID#: _____ Lab Hr: _____

US CUST

ISMC-1A

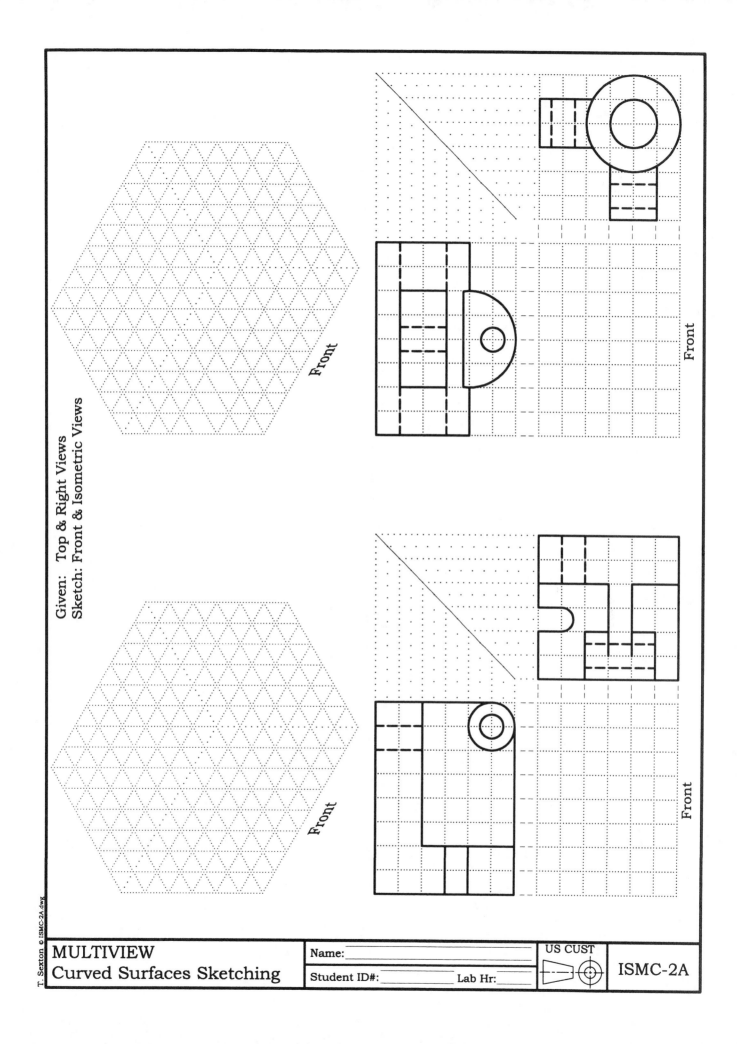

Given: Top & Right Views
Sketch: Front & Isometric Views

Front

Front

Front

Front

MULTIVIEW
Curved Surfaces Sketching

Name:
Student ID#:   Lab Hr:

US CUST

ISMC-2A

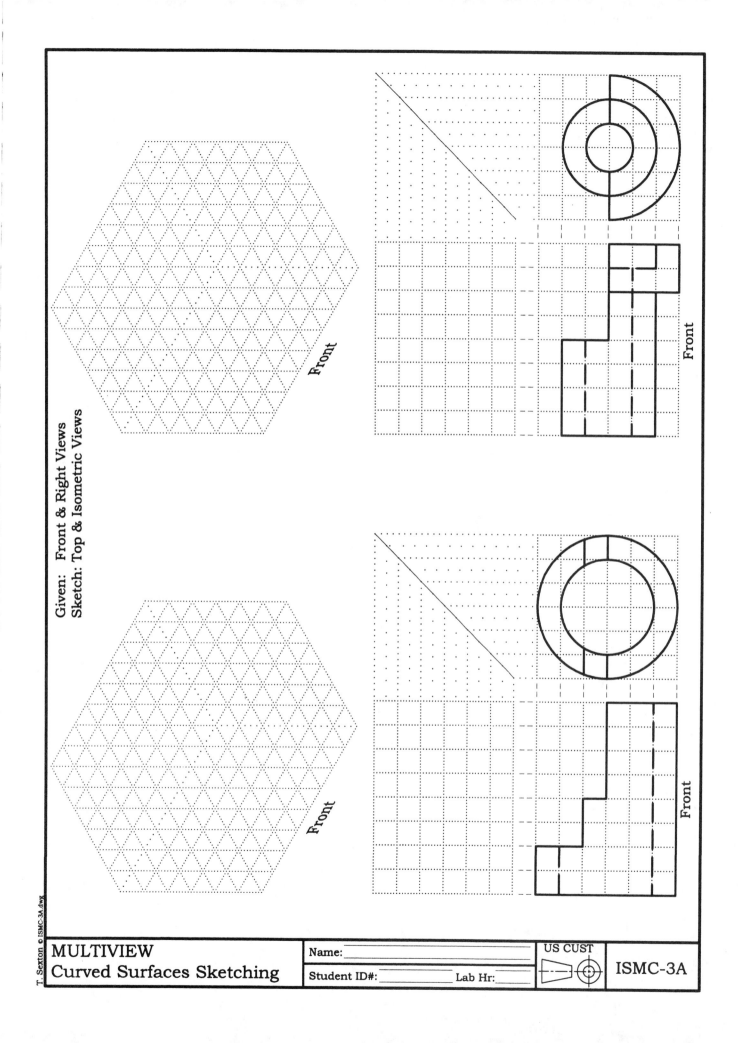

Given: Front & Right Views
Sketch: Top & Isometric Views

Front

Front

Front

Front

MULTIVIEW
Curved Surfaces Sketching

Name:

Student ID#:   Lab Hr:

US CUST

ISMC-3A

T. Sexton © ISMC-3A.dwg

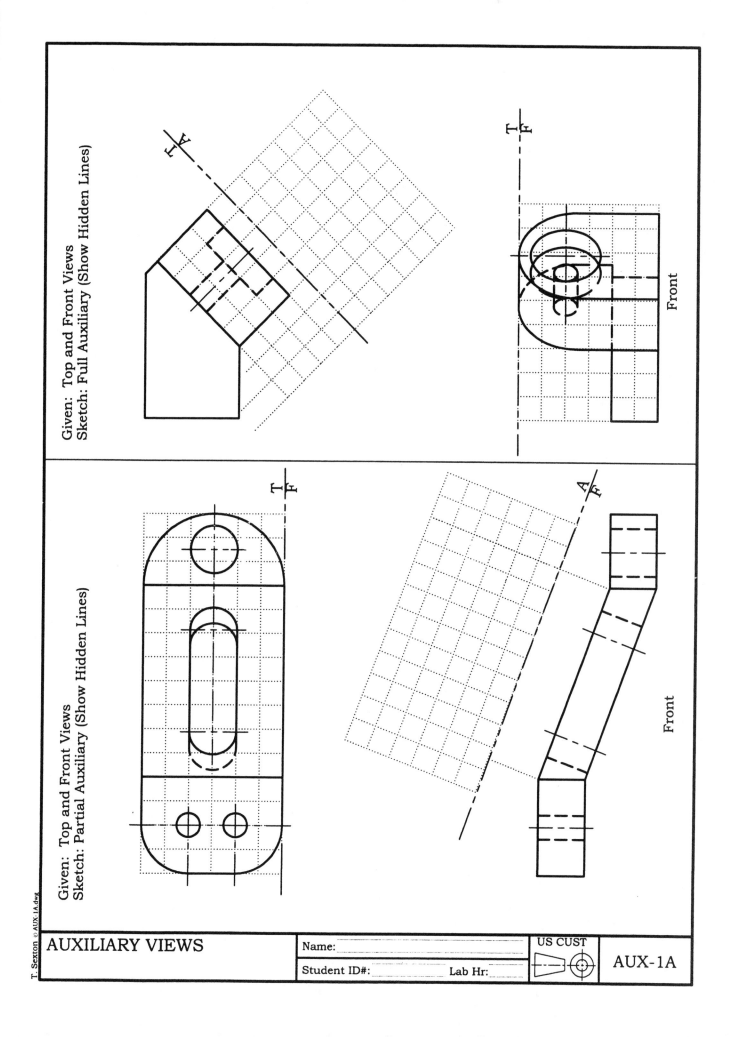

Given: Top and Front Views
Sketch: Full Auxiliary (Show Hidden Lines)

Front

Given: Top and Front Views
Sketch: Partial Auxiliary (Show Hidden Lines)

Front

AUXILIARY VIEWS

Name:

Student ID#:　　　　　　Lab Hr:

US CUST

AUX-1A

T. Sexton ⊘AUX-1A.dwg

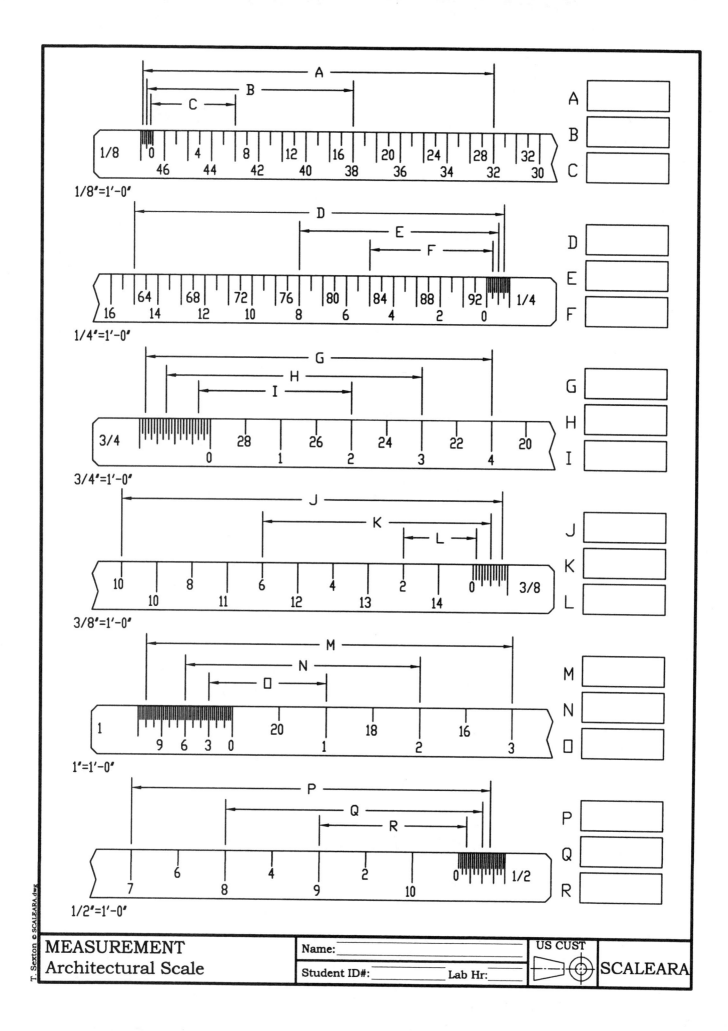

MEASUREMENT
Architectural Scale

Name:

Student ID#:   Lab Hr:

US CUST

SCALEARA

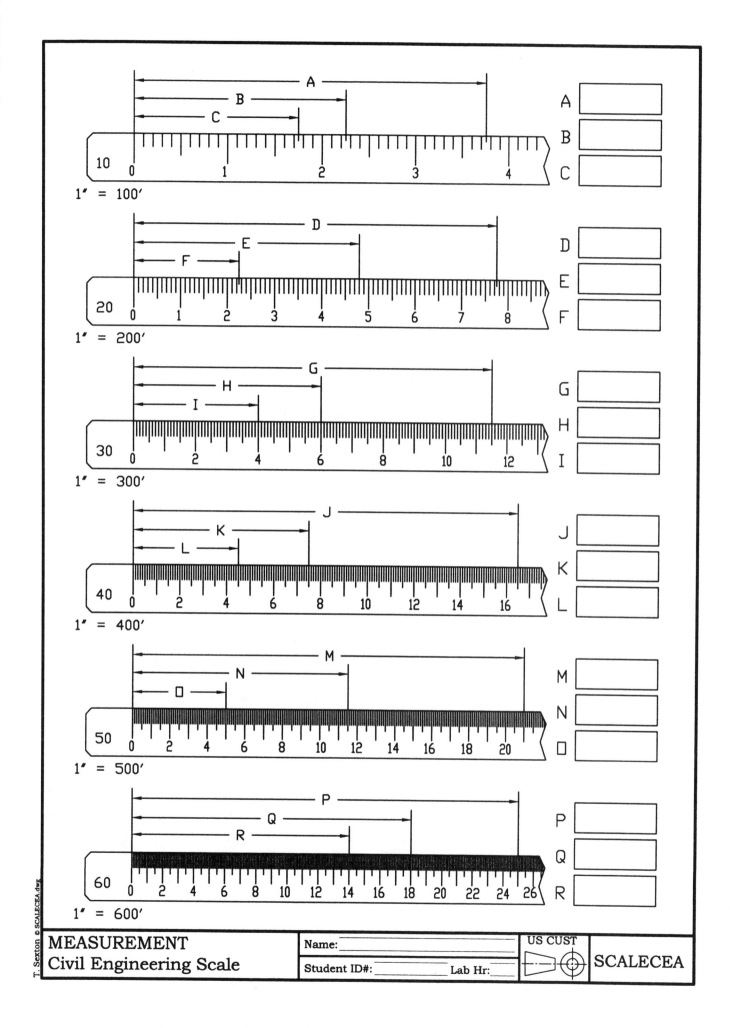

MEASUREMENT
Civil Engineering Scale

Name:

Student ID#: Lab Hr:

US CUST

SCALECEA

T. Sexton ©SCALECEA.dwg

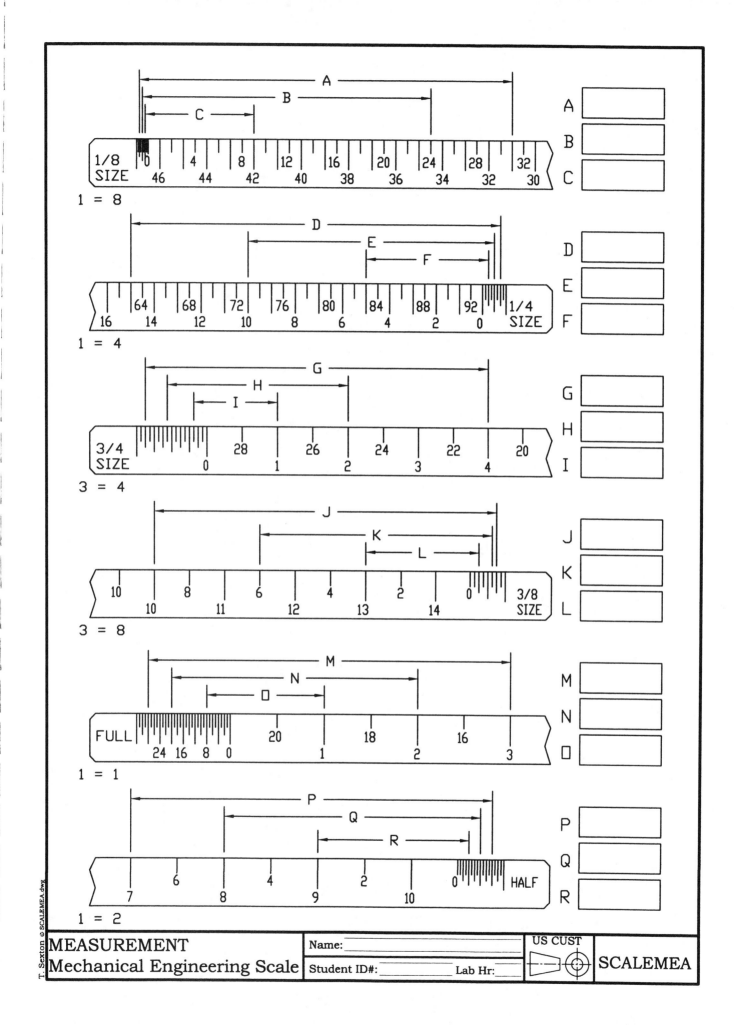

MEASUREMENT
Mechanical Engineering Scale

Name:
Student ID#: _____ Lab Hr: _____

US CUST

SCALEMEA

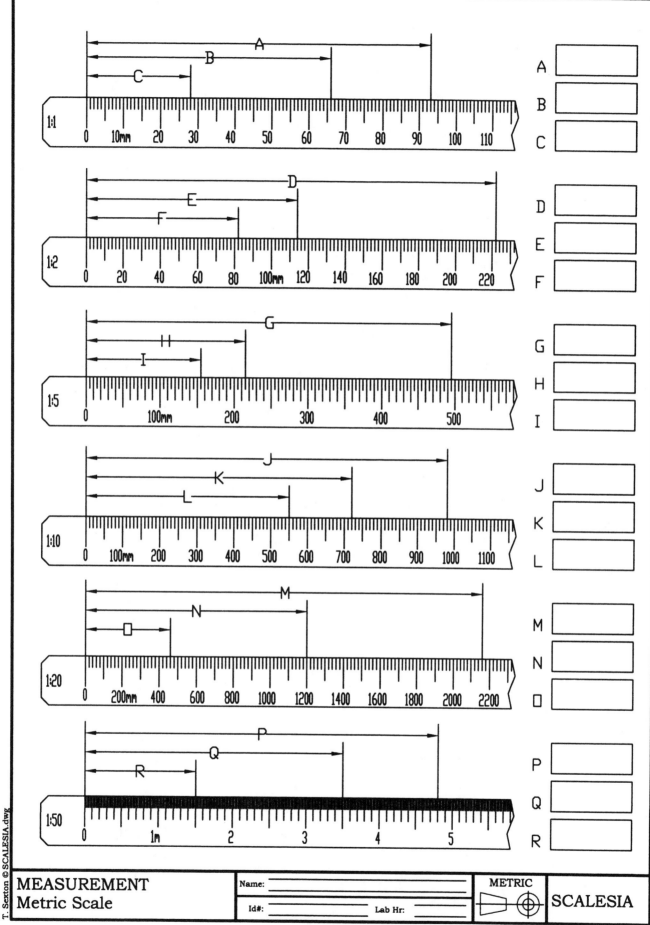

MEASUREMENT
Metric Scale

Name:

Id#:   Lab Hr:

METRIC

SCALESIA

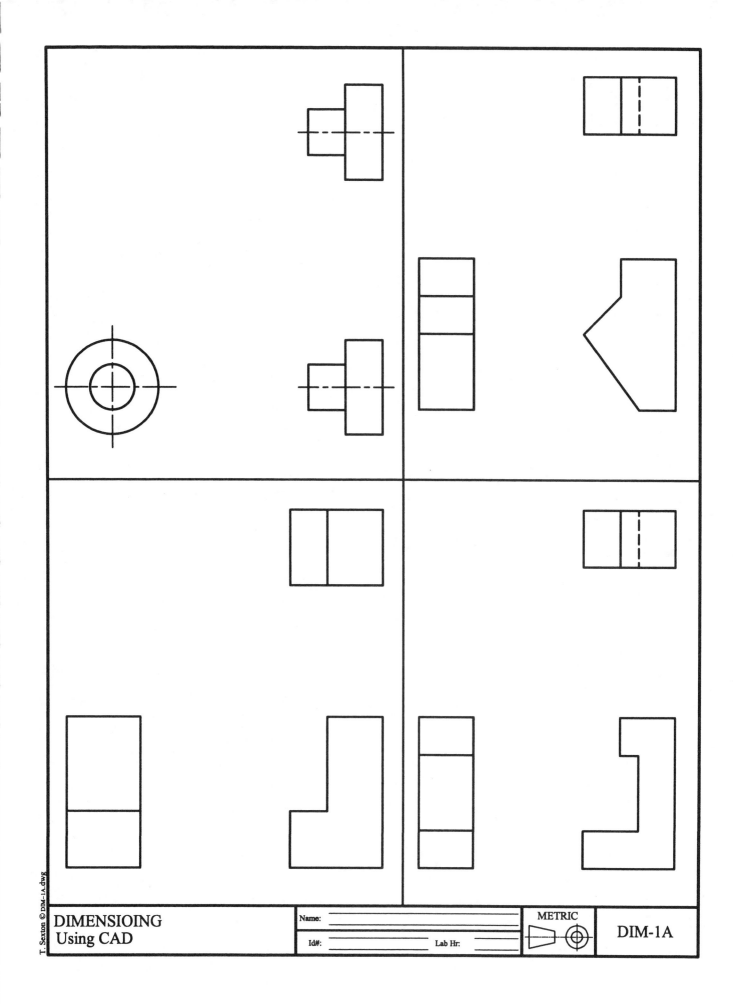

DIMENSIOING
Using CAD

Name:

Id#:   Lab Hr:

METRIC

DIM-1A

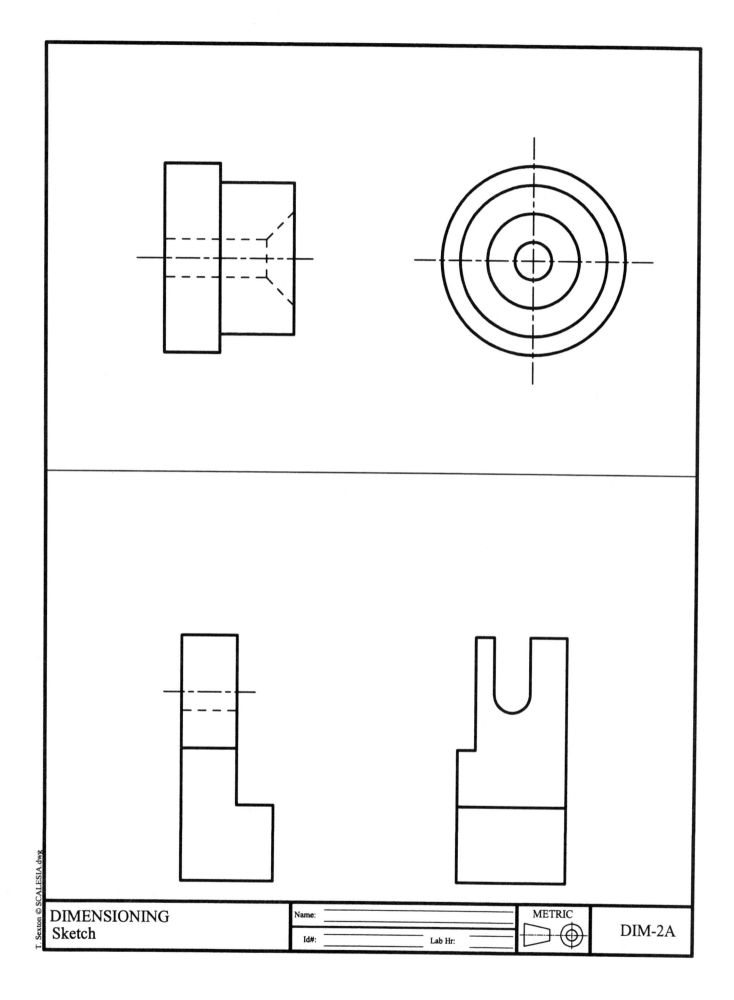

DIMENSIONING
Sketch

Name:

Id#: _____ Lab Hr: _____

METRIC

DIM-2A

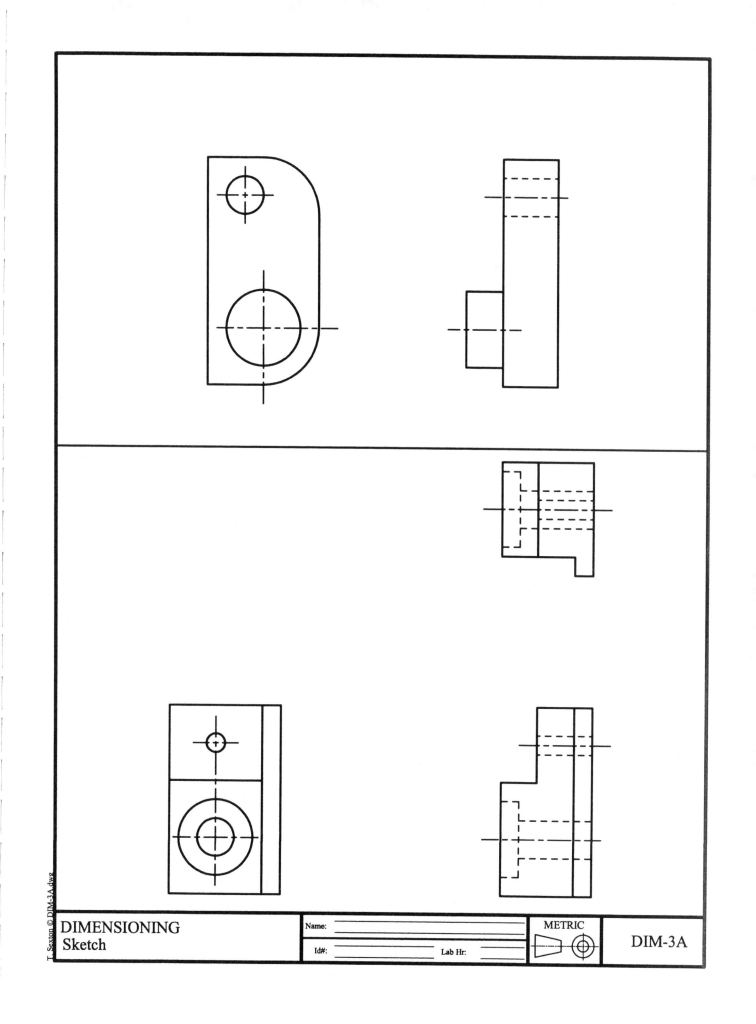

DIMENSIONING
Sketch

Name:

Id#: Lab Hr:

METRIC

DIM-3A

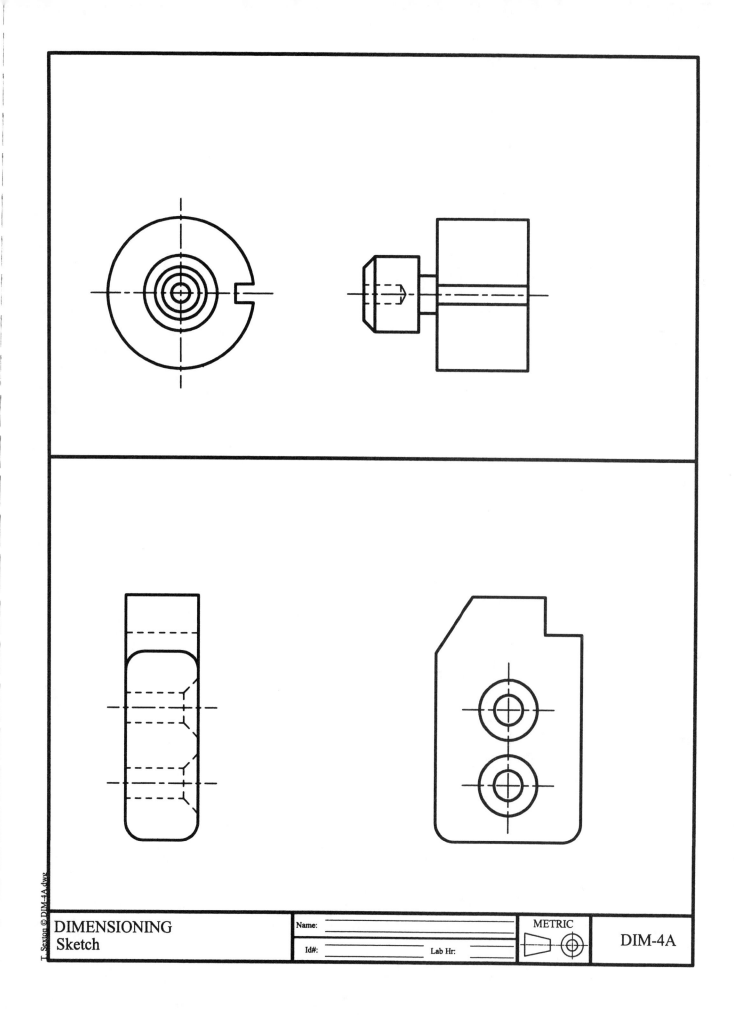

DIMENSIONING
Sketch

Name:

Id#:   Lab Hr:

METRIC

DIM-4A

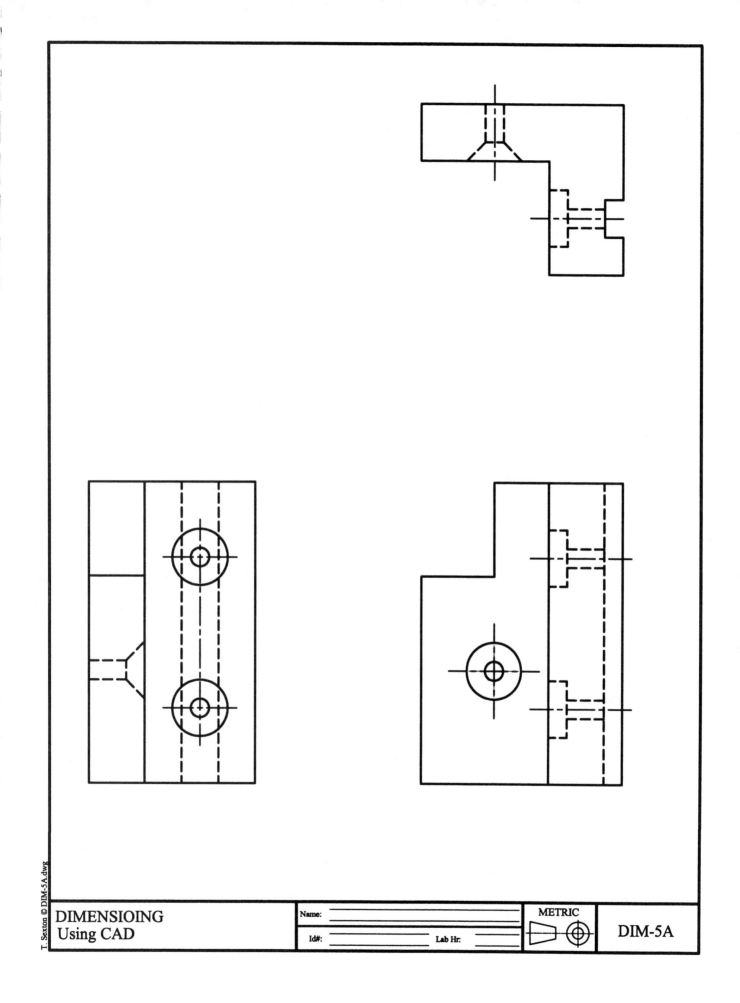

DIMENSIOING
Using CAD

Name:

Id#:   Lab Hr:

METRIC

DIM-5A

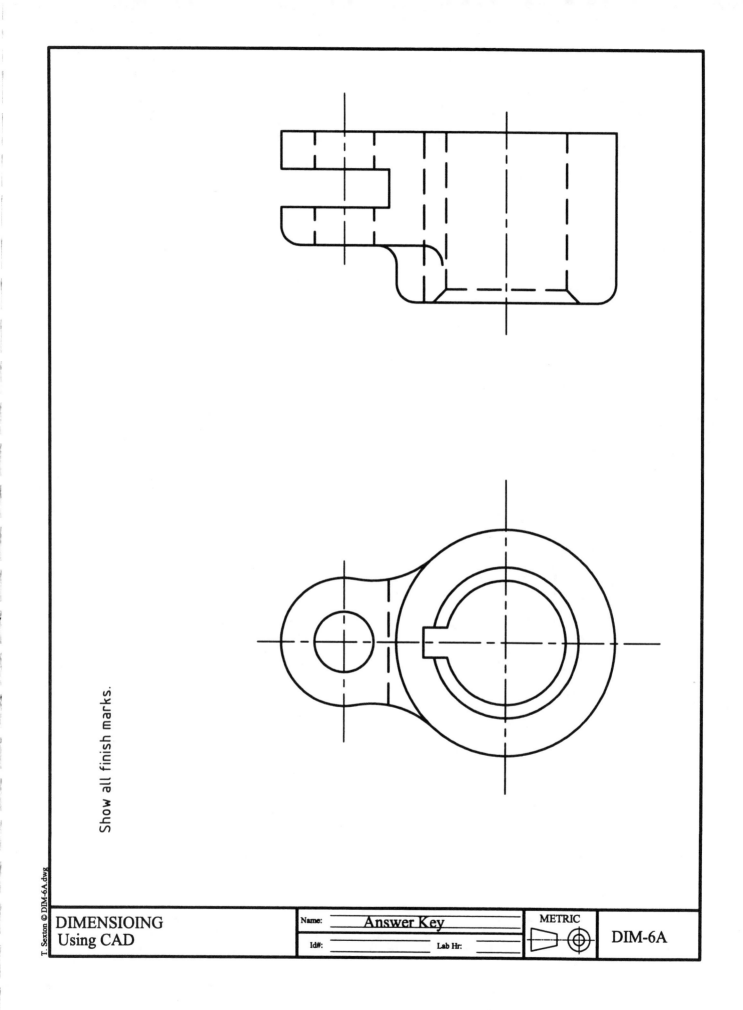

Show all finish marks.

DIMENSIOING Using CAD	Name:	Answer Key		METRIC	DIM-6A
	Id#:		Lab Hr:		

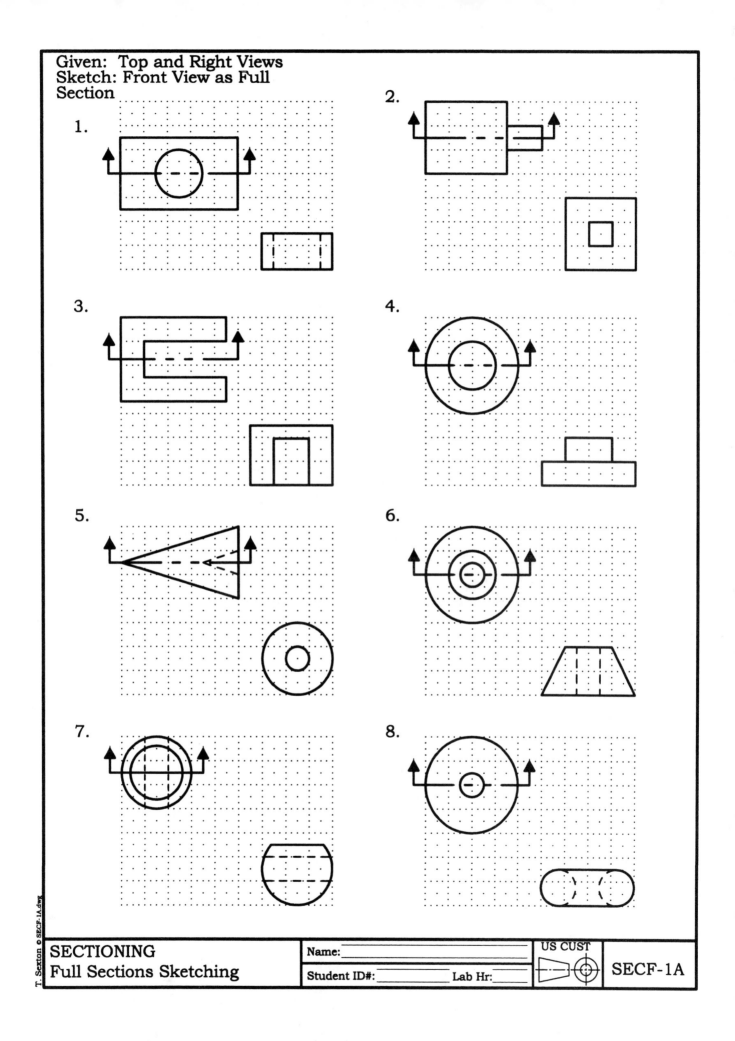

Given: Top and Right Views
Sketch: Front View as Full
Section

1.

2.

3.

4.

5.

6.

7.

8.

SECTIONING
Full Sections Sketching

Name:_____

Student ID#:_____ Lab Hr:_____

US CUST

SECF-1A

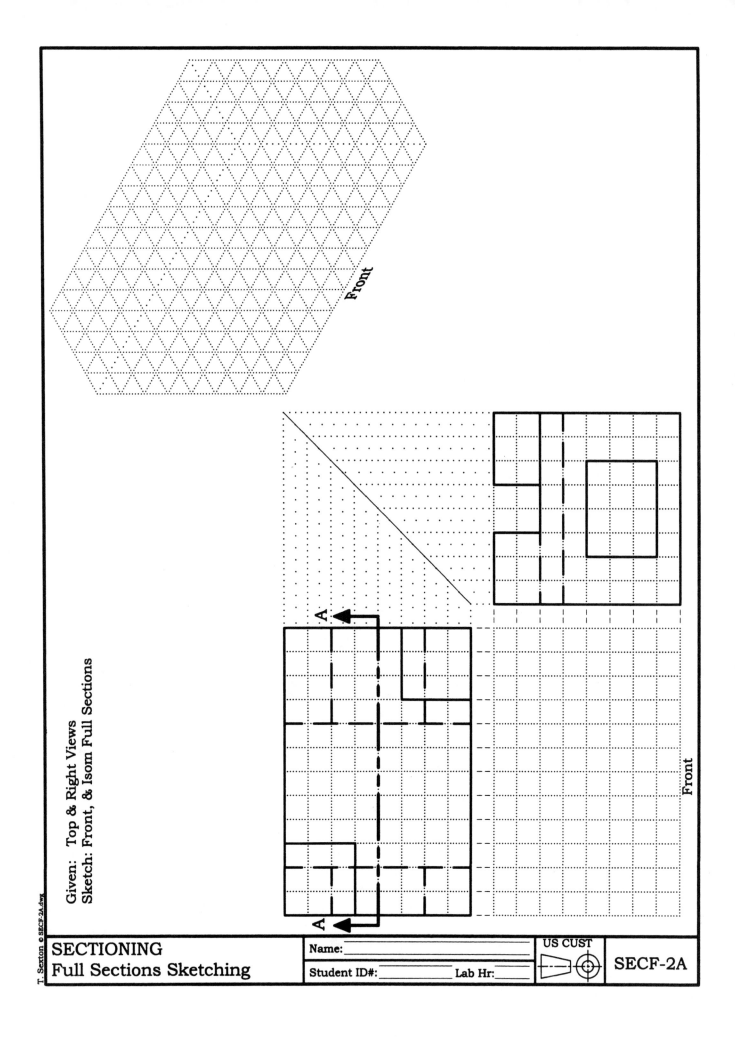

Given: Top & Right Views
Sketch: Front, & Isom Full Sections

Front

Front

A

A

SECTIONING

Full Sections Sketching

Name:_____

Student ID#:_____ Lab Hr:_____

US CUST

SECF-2A

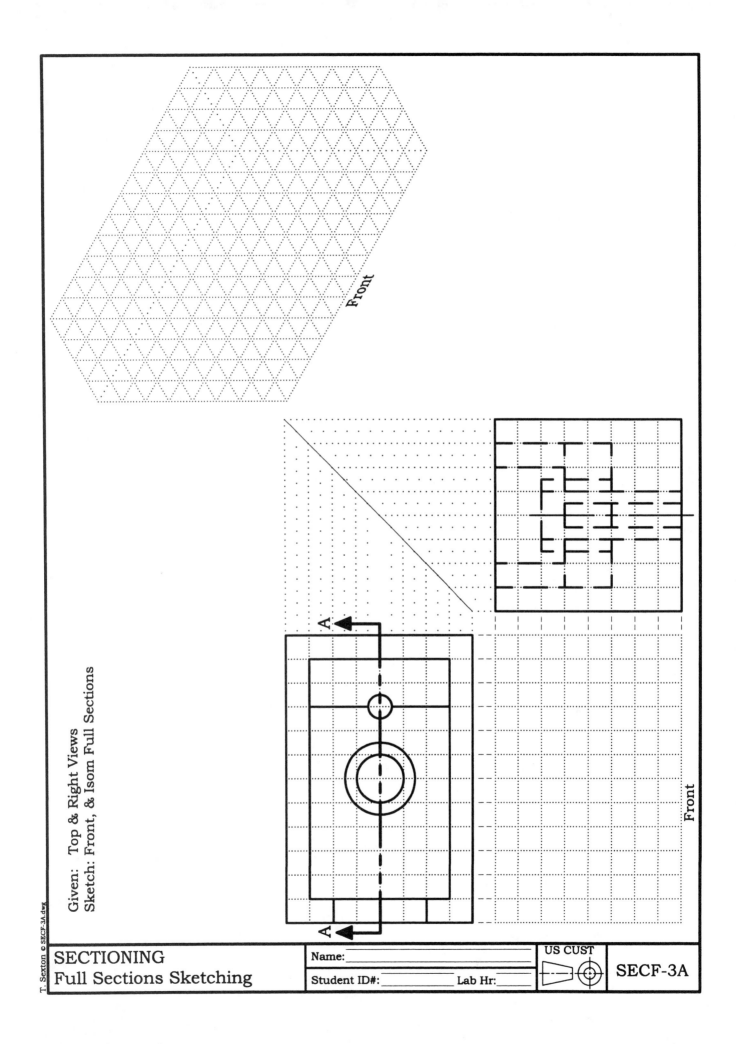

Front

Given: Top & Right Views
Sketch: Front, & Isom Full Sections

Front

SECTIONING
Full Sections Sketching

Name:_____

Student ID#:_____ Lab Hr:_____

US CUST

SECF-3A

A

A

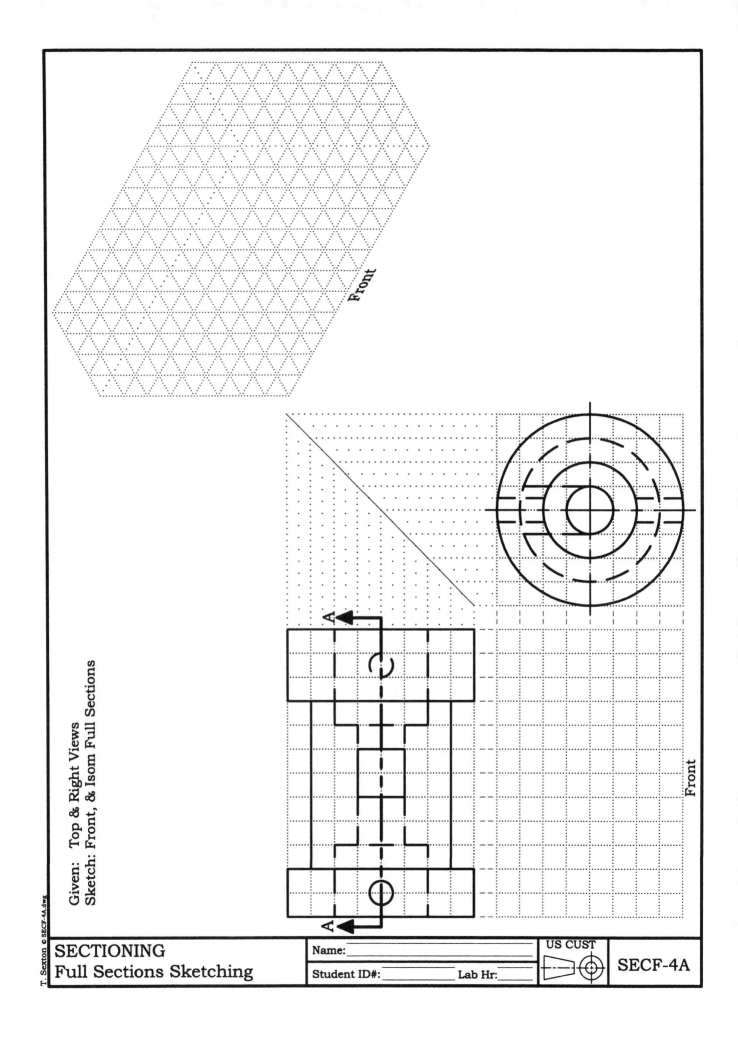

Given: Top & Right Views
Sketch: Front, & Isom Full Sections

Front

Front

A

A

SECTIONING
Full Sections Sketching

Name:

Student ID#: Lab Hr:

US CUST

SECF-4A

T. Sexton © SECF-4A.dwg

Draw the front views as half sections.

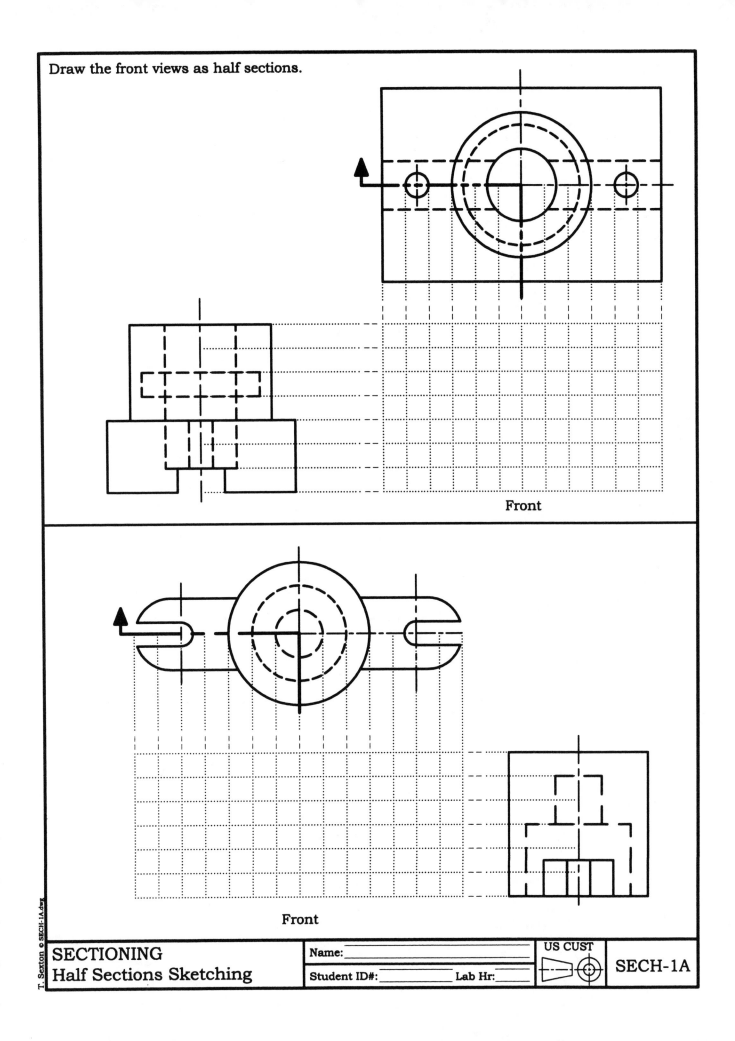

Front

Front

SECTIONING
Half Sections Sketching

Name:

Student ID#: _____ Lab Hr:_____

US CUST

SECH-1A

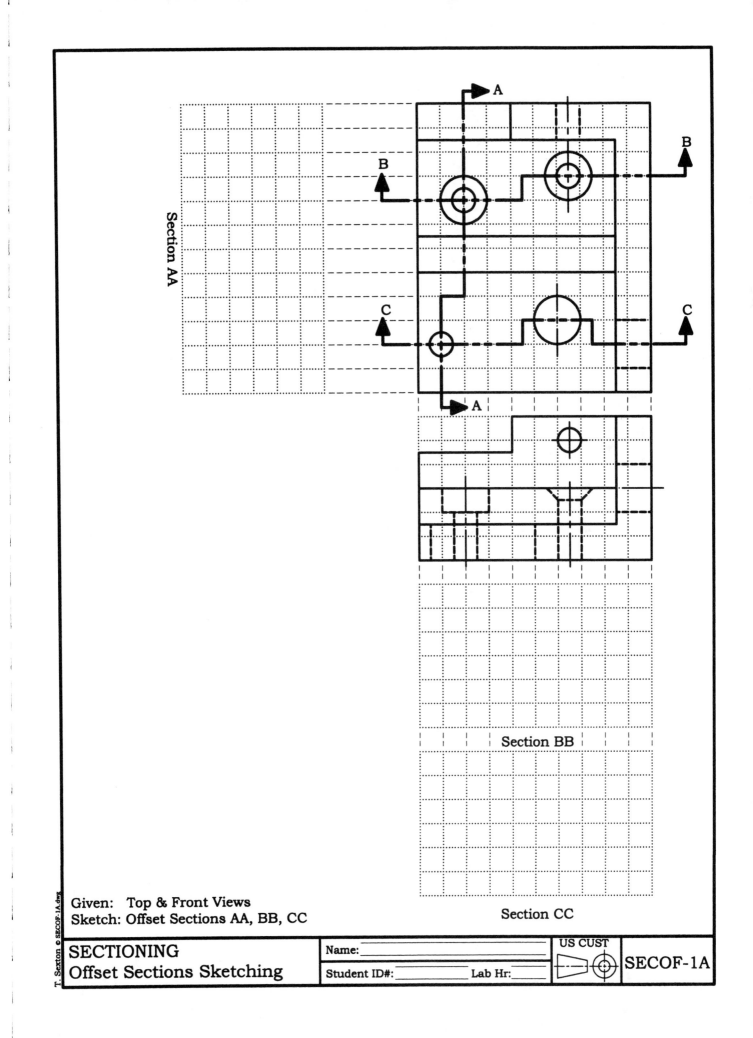

Section AA

A

B

B

C

C

A

Section BB

Section CC

Given: Top & Front Views
Sketch: Offset Sections AA, BB, CC

SECTIONING
Offset Sections Sketching

Name:

Student ID#: _____ Lab Hr:_____

US CUST

SECOF-1A

T. Sexton © SECOF-1A.dwg

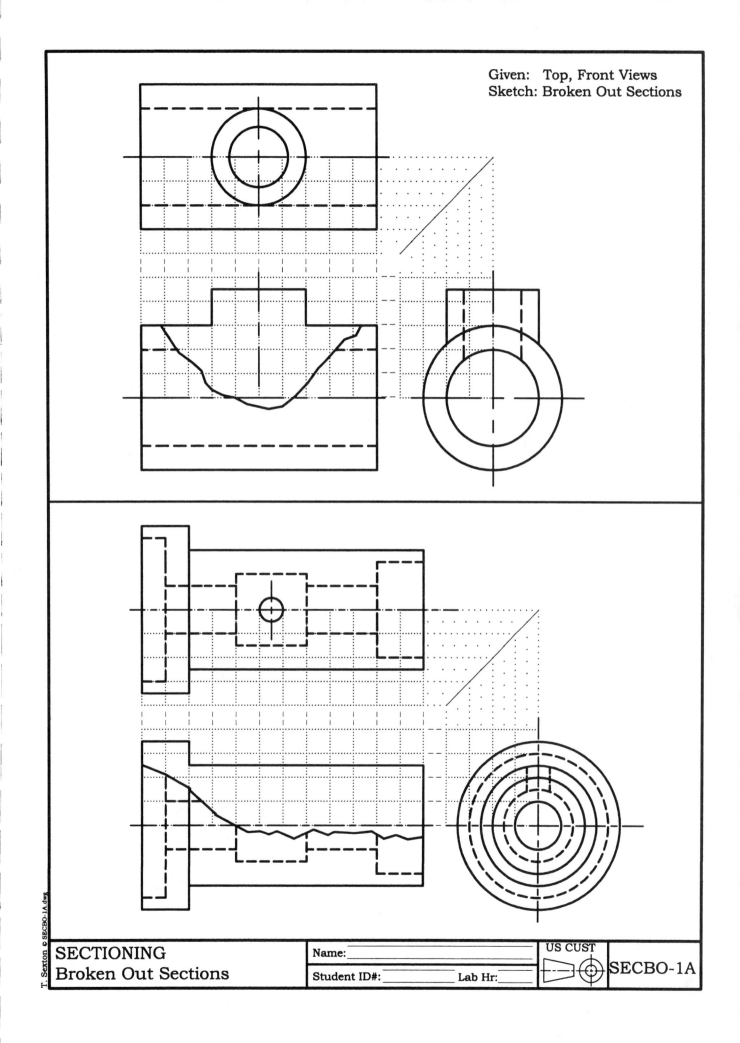

Given: Top, Front Views
Sketch: Broken Out Sections

SECTIONING
Broken Out Sections

Name:
Student ID#: Lab Hr:

US CUST

SECBO-1A

T. Sexton © SECBO-1A.dwg

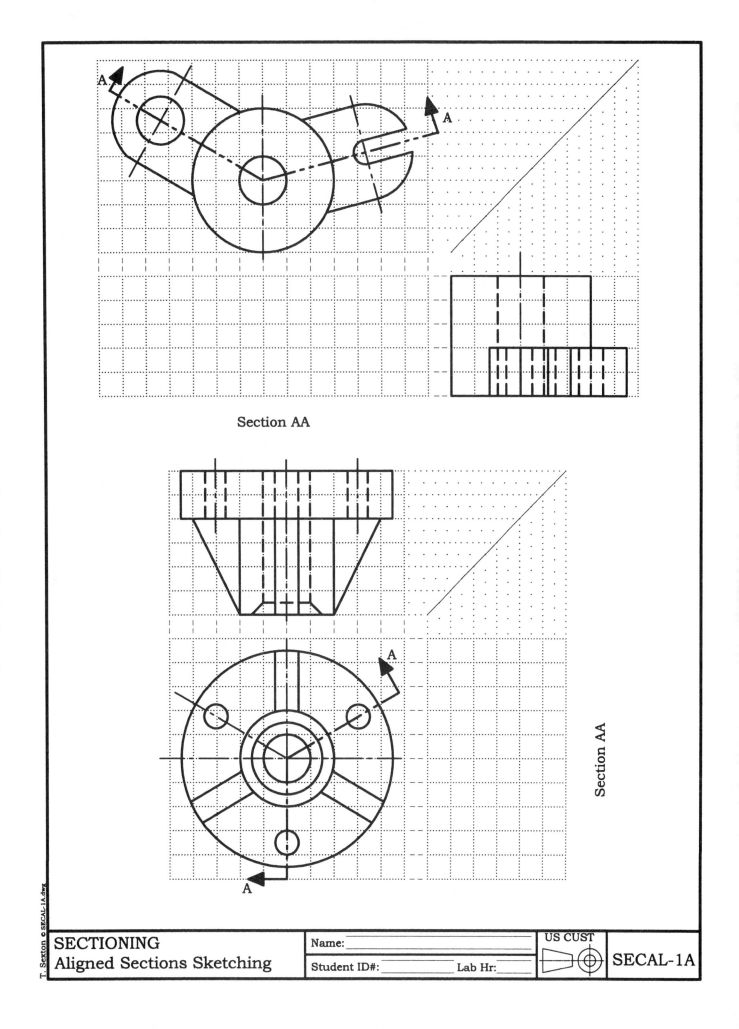

Section AA

Section AA

| SECTIONING | Name: | | US CUST | |
| Aligned Sections Sketching | Student ID#: | Lab Hr: | | SECAL-1A |

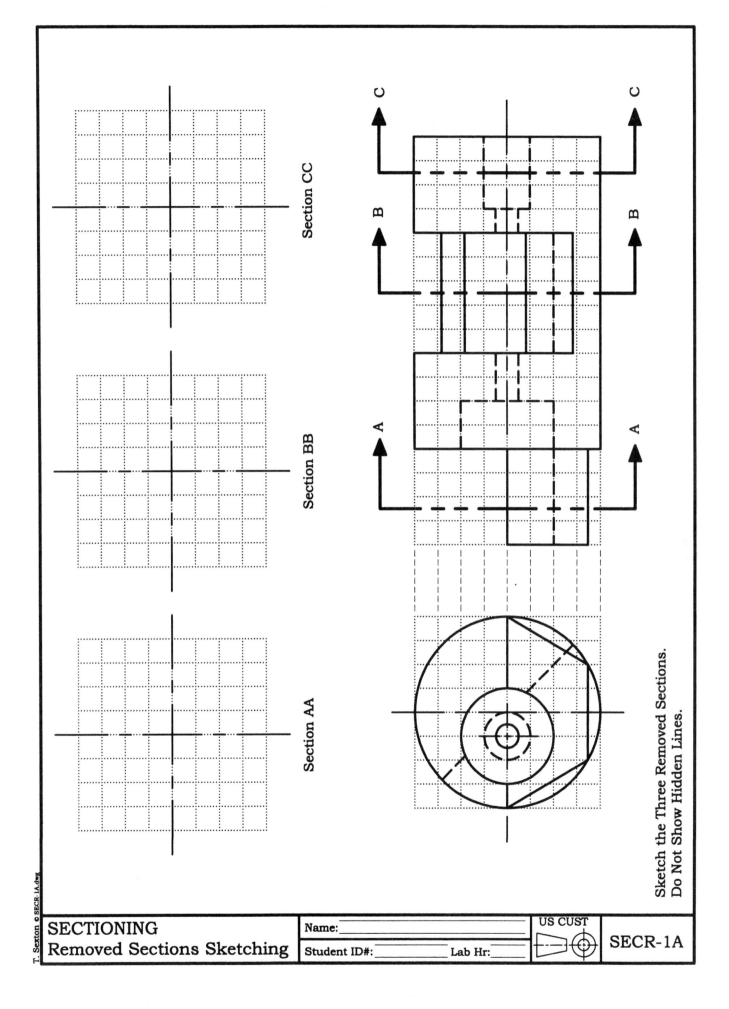

Section CC

Section BB

Section AA

C

B

A

C

B

A

Sketch the Three Removed Sections.
Do Not Show Hidden Lines.

SECTIONING
Removed Sections Sketching

Name:

Student ID#: _____ Lab Hr: _____

US CUST

SECR-1A

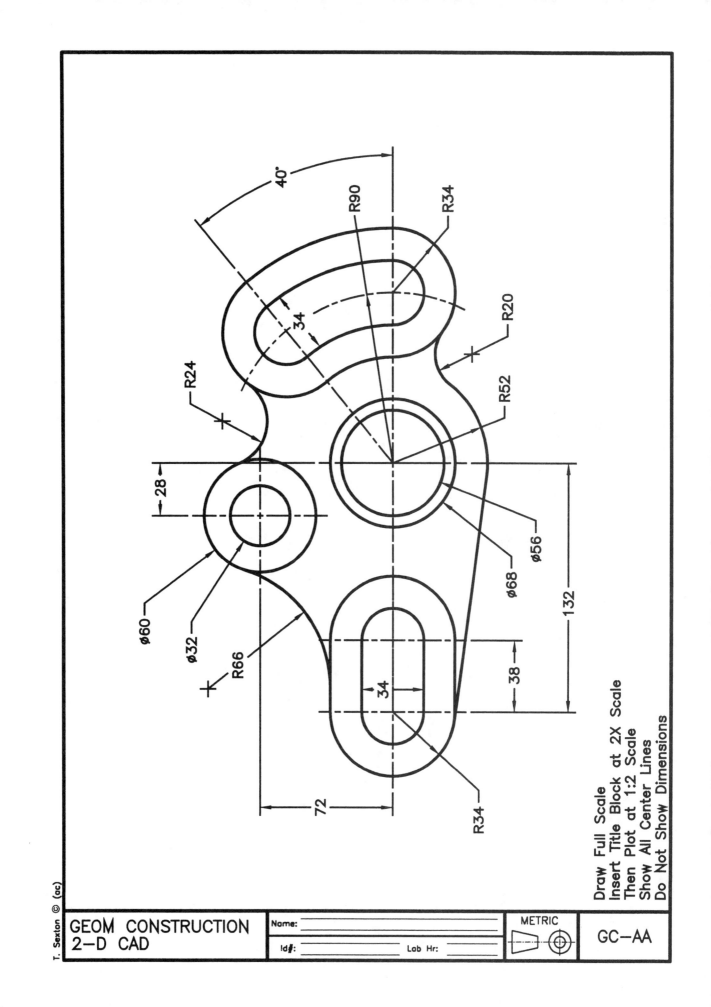

R90

R34

R20

R52

R24

40°

34

R34

Ø56

Ø68

132

Ø60

Ø32

R66

38

34

72

GEOM CONSTRUCTION
2-D CAD

Name:

Id#:   Lab Hr:

METRIC

GC-AA

Draw Full Scale
Insert Title Block at 2X Scale
Then Plot at 1:2 Scale
Show All Center Lines
Do Not Show Dimensions

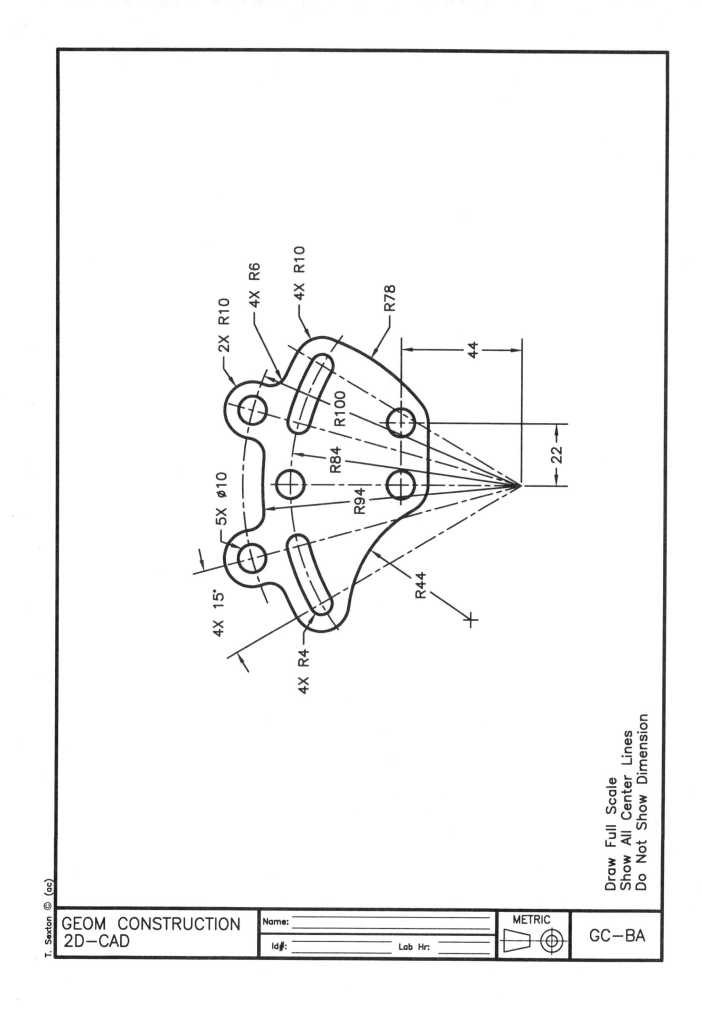

2X R10

4X R6

4X R10

R78

44

R100

R84

22

R94

5X ⌀10

R44

4X 15°

4X R4

GEOM CONSTRUCTION
2D—CAD

Name:

Id#:

Lab Hr:

METRIC

GC—BA

Draw Full Scale
Show All Center Lines
Do Not Show Dimension

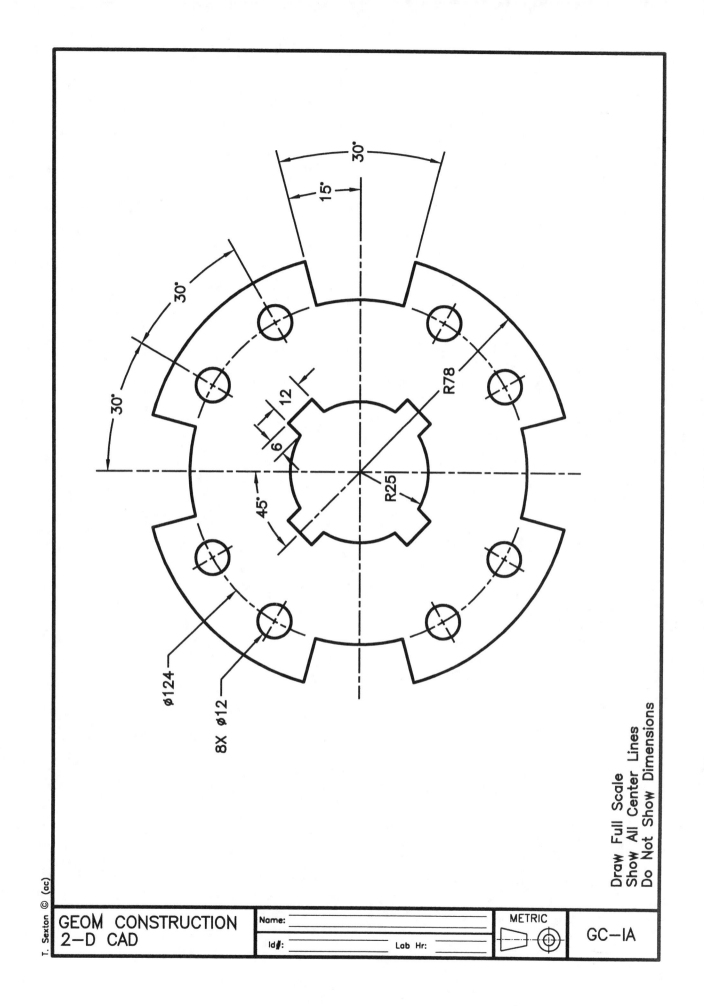

30°

15°

30°

30°

R78

12

6

45°

R25

Ø124

8X Ø12

GEOM CONSTRUCTION
2-D CAD

Name:

Id#:   Lab Hr:

METRIC

GC-IA

Draw Full Scale
Show All Center Lines
Do Not Show Dimensions

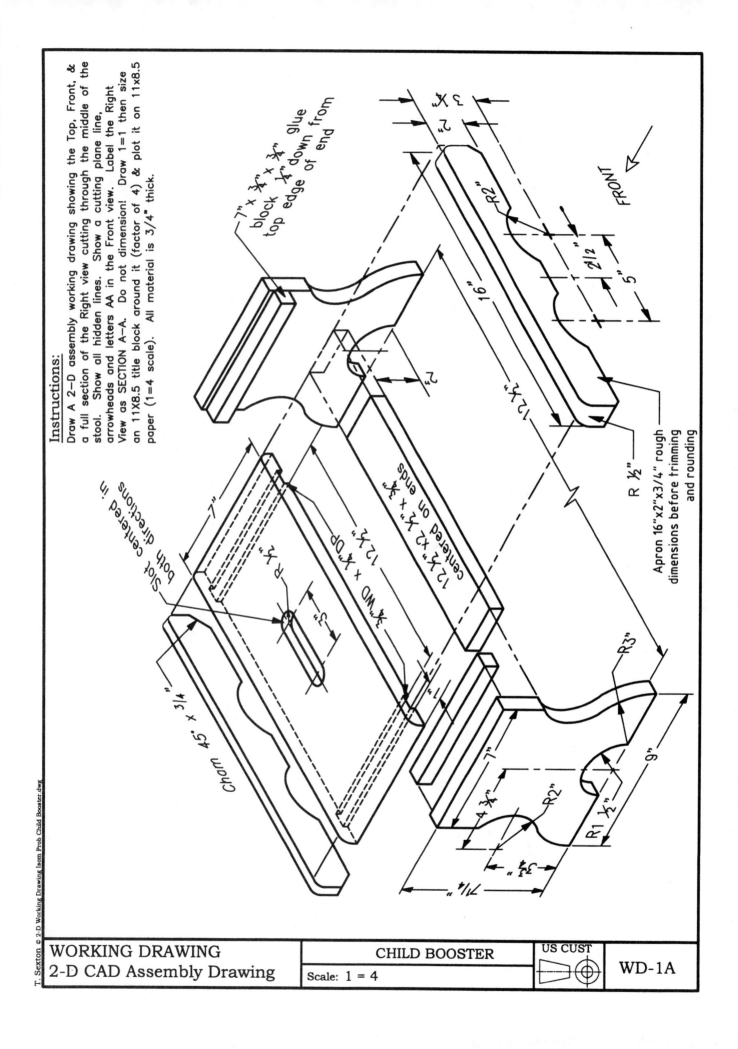

Instructions:

Draw A 2-D assembly working drawing showing the Top, Front, & a full section of the Right view cutting through the middle of the stool. Show all hidden lines. Show a cutting plane line, arrowheads and letters AA in the Front view. Label the Right View as SECTION A-A. Do not dimension! Draw 1=1 then size an 11X8.5 title block around it (factor of 4) & plot it on 11x8.5 paper (1=4 scale). All material is 3/4" thick.

7" x ¾" x ¾" glue block ¼" down from top edge of end

R2"

FRONT

2½"

5"

16"

3¾"

12¾"

R ½"

Apron 16"x2"x3/4" rough dimensions before trimming and rounding

Slot centered in both directions

R ¼"

7"

R ¾" x ½"DP

12¾"

¾"WD x ¾"DP centered on ends
12¾" x 2 ¾" x ¾"

3"

Cham 45° x 3/4

R3"

7"

4 ¾"

R2"

R1 ½"

3 ¾"

7 ¼"

9"

| WORKING DRAWING | CHILD BOOSTER | US CUST | WD-1A |
| 2-D CAD Assembly Drawing | Scale: 1 = 4 | | |

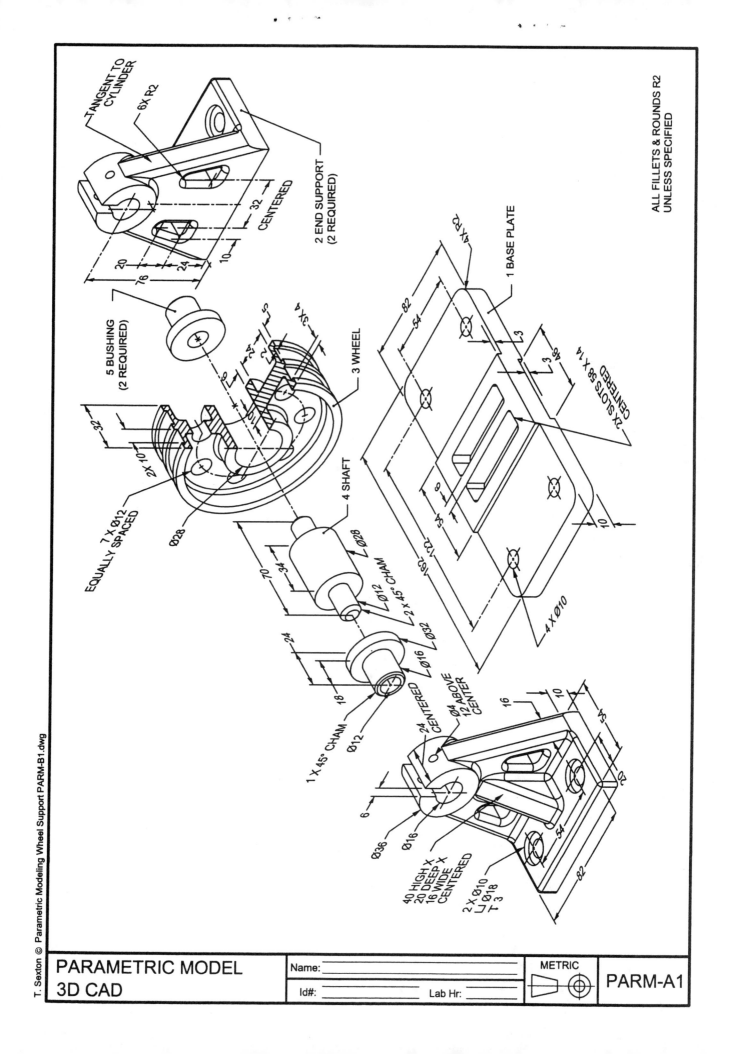

TANGENT TO CYLINDER

6X R2

CENTERED

32

20

24

10

76

2 END SUPPORT
(2 REQUIRED)

5 BUSHING
(2 REQUIRED)

3 WHEEL

4 SHAFT

4X R2

1 BASE PLATE

82

54

54

46

6

3

3

4 X Ø10

162

2X SLOTS 58 X 14
CENTERED

10

EQUALLY
SPACED

7 X Ø12

2X 10

32

Ø28

5

4

3X 4

2

6

10

Ø28

Ø12

2 X 45° CHAM

Ø32

Ø16

70

34

24

18

1 X 45° CHAM

Ø12

Ø4 ABOVE
12 CENTER
CENTER

24 CENTERED

16

10

54

20

6

Ø36

Ø16

40 HIGH X
20 DEEP X
16 WIDE
CENTERED

2 X Ø10
⌴ Ø18
⫧ 3

54

82

ALL FILLETS & ROUNDS R2
UNLESS SPECIFIED

PARAMETRIC MODEL
3D CAD

Name:

Id#: _____ Lab Hr: _____

METRIC

PARM-A1